TECHNICAL REPORT

T0195486

An Examination of the Relationship Between Usage and Operating-and-Support Costs of U.S. Air Force Aircraft

Eric J. Unger

Prepared for the United States Air Force

PROJECT AIR FORCE

The research described in this report was sponsored by the United States Air Force under Contract FA7014-06-C-0001. Further information may be obtained from the Strategic Planning Division, Directorate of Plans, Hq USAF.

Library of Congress Cataloging-in-Publication Data

Unger, Eric J.
    An examination of the relationship between usage and operating-and-support costs of U.S. Air Force aircraft
/ Eric J. Unger.
       p. cm.
    Includes bibliographical references.
    ISBN 978-0-8330-4613-0 (pbk. : alk. paper)
    1. United States. Air Force—Aviation supplies and stores—Costs. 2. Airplanes, Military—United States—
Maintenance and repair—Costs. 3. Airplanes, Military—United States—Parts—Costs. I. Title.

UG1243.U53 2009
358.4'18—dc22
                                                                                    2009002606

Published 2009 by the RAND Corporation
1776 Main Street, P.O. Box 2138, Santa Monica, CA 90407-2138
1200 South Hayes Street, Arlington, VA 22202-5050
4570 Fifth Avenue, Suite 600, Pittsburgh, PA 15213-2665
RAND URL: http://www.rand.org/
To order RAND documents or to obtain additional information, contact
Distribution Services: Telephone: (310) 451-7002;
Fax: (310) 451-6915; Email: order@rand.org

Preface

It is a challenge for U.S. Air Force financial managers to appropriately adjust budgets as flying-hour programs and fleet sizes change. There is a long tradition of using cost-per-flying-hour (CPFH) factors to estimate how budgets should change as flying-hour programs are altered. Unfortunately, there is little empirical support for CPFH factors. A number of prior studies have found them to perform poorly: Actual costs do not necessarily change to the extent predicted by multiplicative application of the CPFH factors.

This report, derived from the Pardee RAND Graduate School (PRGS) dissertation of Air Force Lt Col Eric Unger, systematically examines the empirical relationship between multiple systems' expenditures, flying hours, and fleet sizes. This research suggests a more sophisticated way to think about Air Force costs. Some types of costs vary with flying hours, others with fleet sizes, and still others vary partially with flying hours or fleet sizes. In these "partial" cases, it appears there is a fixed-plus-variable cost structure, with the variable component being less than traditional Air Force CPFH factors.

This research is intended to be of interest to Air Force and other Department of Defense financial management personnel.

RAND Project AIR FORCE

RAND Project AIR FORCE (PAF), a division of the RAND Corporation, is the U.S. Air Force's federally funded research and development center for studies and analyses. PAF provides the Air Force with independent analyses of policy alternatives affecting the development, employment, combat readiness, and support of current and future aerospace forces. Research is conducted in four programs: Force Modernization and Employment; Manpower, Personnel, and Training; Resource Management; and Strategy and Doctrine.

Additional information about PAF is available on our Web site:
http://www.rand.org/paf/

Contents

Figures

Tables

Summary

A central issue in U.S. Air Force budget preparation is how funding levels need to be adjusted as flying hours and fleet sizes change. To address such challenges, the Secretary of the Air Force, Financial Management directorate (SAF/FM) created the expenditure categorization scheme shown in Figure S.1.

In the figure, Cost Analysis Improvement Group (CAIG) expenditure categories are broken into three groups. Those categories labeled *variable cost per flying hour* are assumed to increase or decrease in proportion to flying hours. Those categories labeled *variable cost per TAI* are assumed to increase or decrease in proportion to fleet sizes (total active inventory or

Figure S.1
SAF/FM Breakout of Cost Analysis Improvement Group Level-Two Costs

Variable cost per flying hour	Variable cost per TAI
2.1 POL/energy consumption	1.1 Operations personnel
2.2 Consumables/repair parts	1.2 Maintenance personnel
2.3 Depot-level reparables	1.3 Other mission personnel
2.4 Training munitions/expendable supplies	4.1 Depot maintenance overhaul/rework
2.5 Other unit-level consumption	4.2 Other unit-level consumption
3.3 Intermediate maintenance—transportation	4.3 Depot maintenance engine overhaul
5.1 Interim contractor support	4.4 Depot maintenance other equipment overhaul
5.2 Contract logistics support	
5.3 Other contract support	

Fixed costs
6.1 Support equipment replacement
6.3 Other recurring investment
6.4 Sustaining equipment support
6.5 Software maintenance support
7.1 Personnel support
7.2 Installation support

SOURCE: Lies and Klapper (2007).
NOTES: "Level-one" costs are most aggregated, e.g., mission personnel, intermediate maintenance, depot maintenance. A greater level of detail is found in "level-two" costs, e.g., different types of mission personnel, different types of depot maintenance.
RAND TR594-S.1

tails, TAI). The categories labeled *fixed* are assumed not to vary with flying hours or fleet sizes (though *fixed* does not imply these expenditures could not be reduced or increased).

This report evaluates the validity of Figure S.1 using fiscal year (FY) 1996 to 2006 data on expenditures, flying hours, and fleet sizes for different Air Force aircraft mission designs (MDs) or systems. Our data analysis recommends a somewhat more complicated breakout in which some types of expenses vary partially with flying hours or fleet sizes. For these categories, there appears to be a "fixed-plus-variable" cost structure with incremental costs per flying hour or per tail less than average costs.

Relative to SAF/FM's breakout, we find that a greater proportion of Air Force costs are not variable, especially with respect to flying hours. As a consequence, we are concerned that SAF/FM's approach overbudgets when flying hours increase and underbudgets when they decrease.

Background and Prior Work

CPFH is the primary metric the Air Force uses to create future budgets. Major commands create CPFH factors by mission design series (MDS) or type of aircraft (e.g., F-15C) and multiply the factors by projected flying hours. These projected budget requirements then feed the Air Force's budgetary decisionmaking process.

There is a considerable literature on problems with using flying hours to predict costs.

Hildebrandt and Sze (1990) constructed regression models that relate flying hours to operating and support costs. In general, they found operating and support costs increase less than proportionally with flying hours.

Slay (1995) noted that wartime conditions result in longer sorties and that the number of sorties predicts costs better than the number of flying hours. Sherbrooke (1997) built upon Slay (1995), finding that a disproportionate number of maintenance demands that are unrelated to safety are deferred until the end of a day.

Wallace, Houser, and Lee (2000) found that "removals" (a proxy for cost) are only loosely correlated with flying hours.

Laubacher (2004) examined different forecasting techniques for helicopter budgets. Similarly, Hawkes (2005) studied F-16 costs. Hawkes' primary finding was that last year's CPFH predicts this year's CPFH. Armstrong (2006) studied F-15 CPFH, finding a marginal CPFH would be preferable to average cost.

Data Overview and Estimation Approach

Our analysis is built around two data systems, the Air Force Total Ownership Cost (AFTOC) system and the Reliability and Maintainability Information System (REMIS). AFTOC tabulates expenditures by FY, CAIG cost element category, and weapon system. REMIS tabulates aircraft flying hours and possessed hours.[1] (See pp. 11–12.)

---

[1]   A "possessed hour" refers to the fact that the Air Force *owns* the aircraft, irrespective of whether it is flying or broken.

Our objective is to measure the relationship between expenditures and aircraft flying hours. One could undertake such estimations for each MD separately, but doing so would be hampered by small sample sizes. Instead, we estimated a linear regression of the form

$$\ln(Cost_{it}) = a_i + b * \ln(FH_{it}) + c * Year_t + \varepsilon_{it},$$

where each i is an MD and each t is a year. $Year_t$ is a FY dummy variable. In this estimation structure, each MD gets its own intercept ($\hat{a}_i$) but there is a common $\hat{b}$ that is the most typical empirical relationship between the natural log of flying hours and the natural log of costs. (See pp. 12–14.)

Estimation Results

Using total expenditure data, we estimate an Ln (flying hour) coefficient of 0.56489. This result suggests total spending, on average, increases about 6 percent if a weapon system's flying hours increase 10 percent. (See pp. 15–17.)

In running the analysis of the more-detailed level-one and level-two CAIG categories (see the notes for Figure S.1), the 4.0 (depot maintenance) category shows an unusually large elasticity with respect to flying hours. We believe this finding is spurious and could have been caused by the Air Force changing accounting procedures. (See pp. 17–19.)

For many more-detailed level-two CAIG categories, we find evidence that a category's costs grow with, but not in proportion to, flying hours. The distinct exception is 2.1 (petroleum, oil, and lubricants (POL)/energy consumption) in which, not surprisingly, expenditures closely track flying hours. (See pp. 19–21.)

We also did estimations with both the natural log of flying hours and the natural log of fleet size as independent variables. Such estimations are only feasible because of a post-9/11 increase in flying hours without a commensurate change in fleet sizes. Previously, there was a near-perfect correlation between an MD's fleet size and its flying hours.

The regressions we undertook with both flying hours and fleet sizes as independent variables had mixed findings. Both flying hours and fleet size appear to partially affect total expenditures. Within specific categories, some types of expenditures—e.g., energy consumption—clearly track with flying hours, but others—including maintenance personnel—track with the number of aircraft. Yet other categories have partial, but not proportional, variability with flying hours or fleet size. Results were difficult to interpret for depot maintenance. (See pp. 21–26.)

Figure S.2 presents our alternative Air Force expenditure categorization scheme.

Policy Implications

Our analysis of FY 1996–FY 2006 expenditure and flying hour data suggests possible improvements to SAF/FM's approach. In particular, the "varying with flying hours" and "varying with fleet size" categorizations are too simplistic. In fact, some expenditure categories appear to

Figure S.2
Alternative Air Force Expenditure Categorization Scheme

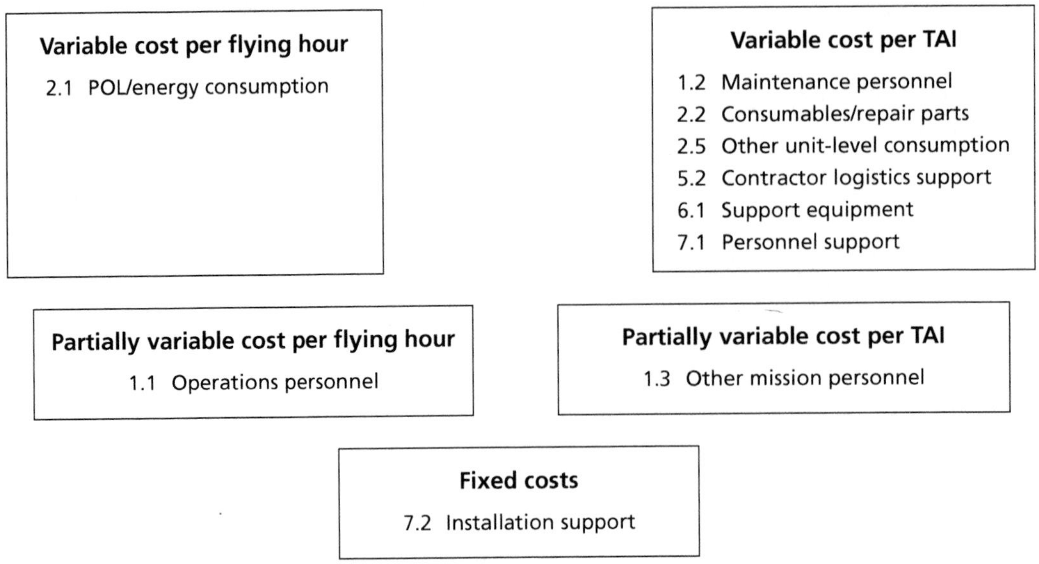

Variable cost per flying hour

2.1 POL/energy consumption

Variable cost per TAI

1.2 Maintenance personnel
2.2 Consumables/repair parts
2.5 Other unit-level consumption
5.2 Contractor logistics support
6.1 Support equipment
7.1 Personnel support

Partially variable cost per flying hour

1.1 Operations personnel

Partially variable cost per TAI

1.3 Other mission personnel

Fixed costs

7.2 Installation support

NOTE: We cannot determine 2.3 + 4.0 (depot maintenance), 2.4 (training munitions), 3.0 (intermediate maintenance), 5.3 (other contract support), 6.3 (other recurring investment), 6.4 (sustaining equipment support), 6.5 (software maintenance), and 6.6 (simulator operations).
RAND TR594-S.2

exhibit fixed-and-variable characteristics, e.g., there is a baseline level of costs in the category and then costs increase with flying hours or fleet size, but not proportionally.

If flying hours are falling, categorizing expenditures as variable when they are actually partially fixed will lead to excessive budget cuts. If flying hours are rising, categorizing expenditures as variable when they are actually partially fixed will lead to excessive budget increases. (See pp. 29–30.)

Acknowledgments

First and foremost, I am grateful for the counseling and patience of my dissertation committee, Ed Keating, Bart Bennett, and Lara Schmidt, for their assistance on the dissertation upon which this technical report is based. Ed Keating, my chair, provided the proper mix of insight and positive (not normative) questions to keep me on the right course. He also holds the distinction of turning around comments quicker than any reviewer in the history of dissertations: I was amazed and impressed. Bart Bennett emphasized the importance of getting my writing started early. That was worthwhile advice, since he helped me expand the scope of my research with deeply intriguing ideas. Lara Schmidt was always the voice of reason in light of the statistics I intended to invent. Her explanations and comments were always clear, accurate, and helpful. I give special thanks to Michael Alles of Rutgers University for his spirited and highly intellectual comments on my work.

Numerous individuals at RAND have helped me accomplish my goals. Natalie Crawford, the consummate mentor, helped me with issues ranging from travel funding to career counseling. Hers was the first voice that welcomed me to RAND—a moment I will never forget. I would also like to thank Michael Kennedy, not only for the funding that allowed me to travel for data collection but also for his time. He and Fred Timson provided valuable input that helped turn the project around at a critical moment.

Adria Jewell, Rodger Madison, and Judy Mele provided computing assistance on this document. Brian Grady and Obaid Younossi helped make my thesis paper into a RAND publication. Laura Baldwin and Cynthia Cook provided comments on the work. Christina Pitcher and Jane Siegel edited the document. Kevin Brancato and Greg Hildebrandt provided helpful and constructive reviews of an earlier draft.

I would like to acknowledge the stellar support I received from the U.S. Air Force. Tom Lies, William "Crash" Lively, Larry Klapper, and Gary McNutt provided access to corporate knowledge that helped me understand the intricacies of AFTOC and operating and support funding. Patrick Armstrong and Eric Hawkes, both Air Force Institute of Technology alumni, injected enthusiasm along with some literature review and data help—a very good combination. Although I was unable to use Programmed Depot Maintenance Scheduling System data for this research, I would like to thank Mark Armstrong for his incredibly professional response to my inquiries.

The cornerstone of my Pardee RAND Graduate School experience was the interaction I had with my very accomplished colleagues. I would like to thank Ying Liu, Yang Lu, and Nailing "Claire" Xia for their help in teaching me Mandarin. Although I only learned two words, I use them both regularly in conversation. I must note that Claire and Yang showed great patience in explaining the nuances of econometric theory to me. However, only Ying had the necessary endurance to discuss microeconomics. Ryan Keefe and Jordan Fischbach set the record at five for attending my dissertation briefings.

Abbreviations

ABIDES	Automated Budget Interactive Data Environment System
AFCAP	Air Force Cost and Performance
AFTOC	Air Force Total Ownership Cost
CAIG	Cost Analysis Improvement Group
CPFH	cost per flying hour
EEIC	Element of Expense and Investment Code
FH	flying hours
FY	fiscal year
MAJCOM	major command
MD	mission design
MDS	mission design series
MSE	mean squared error
O&S	operating and support
PAF	RAND Project AIR FORCE
PEC	Program Element Code
POL	petroleum, oil, and lubricants
PRGS	Pardee RAND Graduate School
REMIS	Reliability and Maintainability Information System
SAF/FM	Secretary of the Air Force, Financial Management directorate
SE	standard error
TAI	total active inventory (number of aircraft)

CHAPTER ONE

Introduction

Developing annual budgets is a major responsibility of the Air Force's financial management community. On a weapon-system-by-weapon-system basis, plans are set forth as to how many aircraft will be in operation and how much each is expected to fly. Financial managers must then make sure that enough, but not too much, funding is allocated to fulfill the desired plan.

While many budgeting activities occur on an annual basis, financial managers may also be called upon to adjust budgets during a fiscal year (FY). For example, a system might be operated more than expected and augmented funding might be required.

Some costs of operating aircraft vary directly with the amount of usage the system gets— fuel costs being a prominent example. Other types of costs, however, do not vary much with usage. For instance, the amount of corrosion-induced maintenance on an aircraft is likely to be a function of an aircraft's age and where it has been stationed, but it has little to do with how much it has been flown.

The Office of the Secretary of Defense's Cost Analysis Improvement Group (CAIG) created a categorization of operating and support (O&S) costs. Costs are assigned to one of seven "level-one" categories as listed in Table 1.1.[1]

The level-one categories are, in turn, broken into "level-two" categories. The categories are additively cumulative, e.g., the five 2.0 (unit-level consumption) level-two categories sum up to equal the 2.0 level-one sum. As shown in Table 1.2, CAIG element 2.0 "unit-level consumption" and 1.0 "mission personnel" have generally been the Air Force's largest level-one O&S categories. The dollar values in all tables are in constant FY 2006 terms using official Office of the Secretary of Defense annual inflation rates. Funding for all Air Force aircraft systems, including unmanned aircraft, are covered in Table 1.2.

Note that the expenditure categories in Table 1.2 are nested, e.g., the 2.1, 2.2, 2.3, 2.4, and 2.5 categories sum (removing rounding error) to the 2.0 total for each FY.

Acknowledging that some costs vary with flying hours (FHs) while others do not, the Secretary of the Air Force, Financial Management directorate (SAF/FM) has developed a categorization scheme that assigns different one- and two-level CAIG groups to "variable with flying hours," "variable with tails," and "fixed." Using the examples noted above, we expect fuel costs to be variable with flying hours, while corrosion maintenance is variable with the number of aircraft in operation. The number of aircraft or, more colloquially, the number of tails, is

---

[1] "Level-one" costs are CAIG's most aggregated breakdown of costs, those seven categories listed in Table 1.1. Within many of the level-one categories, costs are further broken into the more detailed level-two categories. The categories are additively cumulative, i.e., the 2.1–2.5 level-two categories sum up to the 2.0 category total.

Table 1.1
CAIG Cost Element Descriptions

CAIG Element	Description
1.0 Mission personnel	Cost of pay and allowances of officer, enlisted, and civilian personnel required to operate, maintain, and support operational systems
2.0 Unit-level consumption	Includes the cost of fuel and energy resources; operations, maintenance, and support materials consumed at the unit level; stock fund reimbursements for depot-level reparables; operational munitions expended in training; transportation in support of deployed-unit training; temporary duty pay; and other unit-level consumption costs
3.0 Intermediate maintenance	Intermediate maintenance performed external to a unit, including the cost of labor and materials and other costs expended by designated activities/units in support of a primary system and associated support equipment. Includes calibration, repair, and replacement of parts, components, or assemblies and technical assistance
4.0 Depot maintenance	Includes the cost of labor, material, and overhead incurred in performing major overhauls or maintenance on a defense system, its components, and associated support equipment at centralized repair facilities, or on site by depot teams
5.0 Contractor logistical support	Includes the cost of contractor labor, materials, and overhead incurred in providing all or part of the logistics support to a weapon system, subsystem, or associated support equipment. The maintenance is performed by commercial organizations using contractor or government material, equipment, and facilities
6.0 Sustaining support	Includes the cost of replacement support equipment, modification kits, sustaining engineering, software maintenance support, and simulator operations provided for a defense system
7.0 Indirect support	Includes the cost of personnel support for specialty training, permanent changes of station, and medical care. Also includes the costs of relevant host installation services, such as base operating support and real property maintenance

SOURCE: Office of the Secretary of Defense (1992).

formally known as the total active inventory (TAI). The costs of maintaining an Air Force base (7.2, installation support) figure to vary with neither flying hours nor the number of aircraft and so are labeled "fixed."[2]

Figure 1.1 shows how SAF/FM assigns "level-one" CAIG categories to variable-per-flying-hour, variable-per-tail, and fixed costs.

Figure 1.2 presents the same basic categorization, but of "level-two" CAIG categories.

We asked SAF/FM personnel how they derived the breakouts displayed in Figures 1.1 and 1.2. We were told they were based on expert knowledge but were not directly derived from analysis.

This report evaluates the validity of Figures 1.1 and 1.2 using historical data on expenditures, flying hours, and fleet sizes for different Air Force aircraft systems at the mission design

---

[2]   Note that "fixed" does not imply "cannot be reduced." Instead, the meaning is simply that the expenditure level is unrelated to the amount that aircraft fly or the number of aircraft in the fleet. Of course, in the extreme case of the Air Force not having any aircraft nor flying any hours, these "fixed" costs would presumably go away also. So, in reality, we use the term "fixed" to refer to any cost category that is short run, invariant to any reasonable incremental increase or decrease in flying hours or fleet size.

Table 1.2
CAIG Operating and Support Cost Breakout (billions of FY 2006 dollars)

CAIG Element	Costs		
	FY 2004	FY 2005	FY 2006
1.0 Mission personnel	9.44	9.52	9.53
1.1 Operations personnel	2.48	2.55	2.56
1.2 Maintenance personnel	5.79	5.79	5.75
1.3 Other mission personnel	1.17	1.18	1.23
2.0 Unit-level consumption	11.90	12.04	11.30
2.1 POL/energy consumption	5.47	6.18	5.60
2.2 Consumables	1.20	1.13	1.03
2.3 Depot-level reparables	4.47	4.06	3.86
2.4 Training munitions	0.30	0.24	0.35
2.5 Other unit-level consumption	0.46	0.43	0.47
3.0 Intermediate maintenance	0.00	0.00	0.00
4.0 Depot maintenance	2.82	2.88	2.93
4.1 Aircraft depot maintenance	1.87	1.94	1.98
4.3 Engine depot maintenance	0.73	0.70	0.66
4.4 Other depot maintenance	0.21	0.25	0.30
5.0 Contractor logistical support	2.06	2.21	3.24
6.0 Sustaining support	0.67	0.49	0.43
7.0 Indirect support	3.01	2.85	3.03
7.1 Personnel support	0.38	0.37	0.38
7.2 Installation support	2.64	2.48	2.66

SOURCE: Air Force Total Ownership Cost system from the Air Force Total Ownership Cost Web site.

(MD) level.[3] Our data analysis recommends a somewhat more complicated breakout in which some types of expenses vary partially with flying hours or tails. For these categories, there appears to be a "fixed-plus-variable" cost structure, with incremental costs per flying hour, or tail, less than average costs. Relative to SAF/FM's breakout, we find that a greater proportion of Air Force costs are not variable, especially not with respect to flying hours. As a consequence, we are concerned that SAF/FM's approach overbudgets when flying hours increase and underbudgets when they decrease.

The cost metrics discussed in this report (variable cost per flying hour and variable cost per TAI) are used by Air Force cost analysts to compare the total life cycle costs of various

---

[3]   Some mission designs, e.g., F-15s, have mission design series (MDS) nested beneath them, e.g., F-15A, F-15B, F-15C, F-15D, and F-15E. Our analysis is conducted at the MD, not MDS, level.

Figure 1.1
SAF/FM Breakout of CAIG Level-One Costs

Variable cost per flying hour

2.0 Unit-level consumption
3.0 Intermediate maintenance
5.0 Contractor logistical support

Variable cost per TAI

1.0 Mission personnel
4.0 Depot maintenance

Fixed costs

6.0 Sustaining support
7.0 Indirect support

SOURCE: Derived from Lies and Klapper (2007).
RAND TR594-1.1

Figure 1.2
SAF/FM Breakout of CAIG Level-Two Costs

Variable cost per flying hour

2.1 POL/energy consumption
2.2 Consumables/repair parts
2.3 Depot-level reparables
2.4 Training munitions/expendable
     supplies
2.5 Other unit-level consumption
3.3 Intermediate maintenance—
     transportation
5.1 Interim contractor support
5.2 Contract logistics support
5.3 Other contract support

Variable cost per TAI

1.1 Operations personnel
1.2 Maintenance personnel
1.3 Other mission personnel
4.1 Depot maintenance overhaul/
     rework
4.2 Other unit-level consumption
4.3 Depot maintenance engine
     overhaul
4.4 Depot maintenance other
     equipment overhaul

Fixed costs

6.1 Support equipment replacement
6.3 Other recurring investment
6.4 Sustaining equipment support
6.5 Software maintenance support
7.1 Personnel support
7.2 Installation support

SOURCE: Lies and Klapper (2007).
RAND TR594-1.2

force structure mixes. These cost metrics include contractor logistics support and depot maintenance, whose budgets are not built using factors. The factor-driven flying hour budgets cover fuel, consumables, and depot-level reparables only. If the improvements discussed in this report are implemented, then the Air Force will have much more accurate costs for building force structure life cycle trade-off models.

The rest of this report is structured as follows: Chapter Two discusses prior research on cost per flying hour (CPFH) calculations, i.e., the practice of multiplying projected flying hours by a cost-per-hour factor in certain segments of the budgetary process. A number of previous studies have critiqued CPFH approaches. Some of those studies focus on only one or a few weapon systems. By contrast, in this report, we look across Air Force MDs and estimate general, historical relationships between expenditure levels and flying hours. Chapter Three presents an overview of the Air Force Total Ownership Cost (AFTOC) and Reliability and Maintainability Information System (REMIS) data we use.

Chapter Four presents our main estimation results, both aggregating CAIG categories and looking specifically at each major level-one and level-two category. We conclude Chapter Four by suggesting a different CAIG breakout. In Chapter Five, we discuss the policy implications of our findings. We note that we believe that current Air Force budgeting approaches overestimate funding needs when flying hours are increasing and underestimate needs when flying hours are decreasing.

CHAPTER TWO

Background and Prior Work

This chapter provides background information on how the Air Force uses CPFH factors. Then we discuss the existing literature, much of which is unfavorable to the CPFH approach.

Background

We start with an explanation of the CPFH budget formulation process. The CPFH metric, referenced in U.S. Air Force (1994), is the primary metric the Air Force uses to create future budgets from historical costs and flying hours. The creation of individual, mission design series (MDS)–specific CPFH factors is a multistep process that involves many stages of input and review (Rose, 1997).

Figure 2.1 shows a simplified representation of the process by which the Air Force translates projected flying hours and MDS-specific CPFH factors into initial estimates of budgets for consumables, spares, and fuel costs. Using input from the aircraft operators, major command (MAJCOM) analysts create a CPFH factor for a given MDS such as the F-15E. MAJCOM CPFH estimates are subject to reviews by numerous Air Force organizations. For example, the Air Force Cost Analysis Agency performs a "test of reasonableness" on such CPFH estimates.

The resulting CPFH factor is used to adjust budgets when projected flying hours change. There are MDS-specific CPFH factors for spare parts, aviation fuel, and consumables; the Air Force budgets for these categories via Element of Expense and Investment Codes (EEICs) (U.S. Air Force, 1999). The spare parts category includes EEIC 644: flying reparable aircraft parts—those parts that can be repaired, usually by a depot, and are used in direct support of the Air Force's flying hour program (e.g., aircraft engines). Aviation fuel includes EEIC 699 (aviation fuel) and EEIC 693 (nonflying aviation fuel, which is used for engine repair activities). The consumables category includes EEIC 609 (aircraft parts that are not repaired, such as nuts and bolts, but are purchased through base supply) and EEIC 61952 (consumable aircraft

Figure 2.1
Simplified CPFH Metric

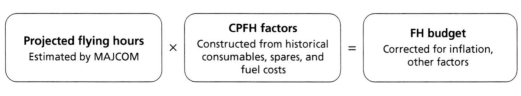

RAND *TR594-2.1*

parts purchased outside base supply).[1] MAJCOM analysts use five years of data to compute fuel requirements, with the other categories computed from two years of data.

Once the approved MDS estimates are completed, the MAJCOM aggregates the budget requirements and submits them to the central Air Force budget system, which is called the Automated Budget Interactive Data Environment System (ABIDES). Air Force decisionmakers use requirement information in ABIDES to request budget authorizations from Congress (U.S. Air Force, 1999).

Prior Work

In this section, we discuss prior work that bears on CPFH issues.

Hildebrandt and Sze (1990) constructed regression models that relate multiple systems' flying hours to several different subelements of O&S costs. They used a log-log regression specification so their flying-hour coefficients can be interpreted as elasticities. Our estimations discussed below are structured similarly.

In general, Hildebrandt and Sze (1990) found that O&S costs increase less than proportionally with flying hours, e.g., if flying hours double, costs will increase but not double.

For total O&S cost per aircraft, Hildebrandt and Sze (1990) found that a 10 percent increase in flying hours per aircraft results in a 6.2 percent increase in cost. As discussed below, with more and newer data, we nevertheless estimate a very similar flying hour–total cost elasticity.

Slay (1995) focused on the differences between wartime and peacetime operation of fighter aircraft. He showed that peacetime-derived flying hour–spare parts relationships grossly overpredict spare parts needs during a wartime flying surge, e.g., the first Gulf War. Slay (1995) argued that average sortie durations increase during wartime and that the number of sorties flown, not the number of flying hours, drives demand for spare parts.

Slay's Logistics Management Institute colleague Sherbrooke (1997) refined Slay's work by presenting regression models for 24 MDS that related spares demand to the sortie number of the day, mission type, location, and sortie duration. One of Sherbrooke's intriguing findings is that demand for spares after the completion of the last sortie in a day is disproportionate: It appears that flight maintenance problems that are not safety related are deferred until the end of the day.

Wallace, Houser, and Lee (2000) focused on C-5Bs during Operation Desert Storm and C-17s, KC-135s, and F-16Cs during the late 1990s' operations in Kosovo. They found that there are other factors that contribute to aircraft maintenance. In addition to flying hours, they found that the critical parameters to forecasting maintenance needs are ground days, cold cycles (engine start and shut down—the number of times an engine is started cold), and warm cycles (pairs of landings and takeoffs during a sortie in which the engines are not shut down). As long as there are small changes in the flying hour program, the proportional CPFH model performs well. However, a more complex model that Wallace, Houser, and Lee developed outperforms the proportional CPFH model during contingency surges, in which flying hours increase dramatically, but landings and maintenance needs do not.

---

[1]   Maj Dane Cooper, Air Force Cost Analysis Agency, Crystal City, Arlington, Va., email correspondence, May 2007.

Several recent Air Force Institute of Technology theses have studied CPFH issues on a system-specific basis. Laubacher (2004) examined three separate forecasting techniques for the Air Force's MH-53J/M, HH-60G, and UH-1N helicopters, with the goal of reducing the differences between forecasted MAJCOM budgets and actual expenses. Hawkes (2005) studied F-16s; he found that last year's CPFH is a good predictor of this year's CPFH. Armstrong (2006) studied F-15 data to estimate a marginal or incremental, rather than average, CPFH.

Next, we discuss our estimation approach. While the Air Force Institute of Technology work focused on specific systems, our approach is more similar to Hildebrandt and Sze (1990) in the sense of trying to estimate more general or typical relationships between costs and independent variables such as flying hours, average fleet age, and fleet sizes.

Our broad findings are in accord with all of the aforementioned prior work in the sense of suggesting that the Air Force can improve the accuracy of its budgeting process by moving beyond average CPFH factors. These factors are acceptable for some types of expenses—e.g., fuel costs—but perform quite poorly for others.

CHAPTER THREE

Data Overview and Estimation Approach

Our analysis is built around AFTOC and REMIS. Therefore, we start with an overview of the information provided by these data systems.

AFTOC tabulates expenditures by FY, CAIG cost-element category, and weapon system. AFTOC data have been produced from underlying Program Element Code (PEC) expenditure records. PECs are very complicated groupings that can change over time. One of the values of AFTOC is that it presents multiple years' expenditure information in a consistent format, sparing the AFTOC user a requirement to track (or even be familiar with) PECs.

AFTOC provides expenditure data both in then-year and inflation-adjusted terms. Inflation adjustments are undertaken using official Office of the Secretary of Defense annual inflation rates. We use inflation-adjusted expenditure data throughout our analysis.

AFTOC presents expenditure data both by MD and MDS. Unfortunately, AFTOC's MDS-level expenditures are often just flying hour–based allocations of MD-level expenditures. Many of the Air Force's PECs do not distinguish across MDS, so break down by AFTOC's MDS level requires allocation assumptions. Aircraft inventory data are used to allocate military personnel costs, while other categories are allocated by flying hour (e.g., F-16 consumable expenditures are divided among the F-16's variants in proportion to annual flying hours). Such allocation is, perhaps, unavoidable from AFTOC's perspective. In our view, however, this phenomenon adds uncertainty to the AFTOC MDS totals, so we limit ourselves to MD-level analysis. All results presented herein are based on MD-level, rather than MDS-level, analysis.

According to the Air Force Web site, "REMIS provides authoritative information on weapon system availability, reliability and maintainability, capability, utilization, and configuration." For this analysis, we make use of only a small fraction of REMIS's information, its tabulation of aircraft flying hours, and its "possessed hours" (described below).

REMIS tabulates monthly flying hours by aircraft (at the specific tail-number level), but we aggregate the flying-hour data to the annual MD level since AFTOC expenditure data are annual and, as discussed above, we believe that the expenditure data are less affected by allocation rules at the MD level. REMIS does not directly break up flying hours between deployed versus nondeployed, or peacetime versus contingency. Our conceptualization is that there is a relatively stable level of peacetime flying hours, so observed changes in flying hours might be ascribed to contingencies increasing or decreasing. But, in fact, we simply observe changes in total flying hours without direct explanation of why they changed.

Possessed hours are a tabulation of hours that a plane is owned by the Air Force. Possessed hours are used to compute average fleet size in a year, i.e., the sum of possessed hours divided by 8,760 (in a non–leap year). We do not differentiate across different possession purpose codes, e.g., whether an aircraft is possessed by operating commands or by the depot system.

Table 3.1 illustrates our combined REMIS-AFTOC data using the B-1 as an example. The flying-hours and TAI columns are derived from REMIS. The CAIG-total column comes from AFTOC. We also display the CAIG 2.0 (unit-level consumption) total, illustrating how the AFTOC data can be analyzed at different levels of granularity. The CAIG 2.0 total is, by definition, included in the CAIG total. All CAIG dollar values, as with the values in all of our tables, are in constant FY 2006 terms.

Estimation Approach

A major objective of this research is to measure the relationship between expenditures and aircraft flying hours.

A logical way to estimate such a relationship would be MD-level linear regression, e.g., regress B-1 AFTOC expenditures on B-1 flying hours. Figure 3.1 shows the result of this regression.

The regression has an estimated intercept term of about $790 million, with a slope of about $15,000 per flying hour. By contrast, the 1996–2006 average cost per B-1 flying hour was about $48,000. This single-system result suggests considerable fixed costs with a much lower incremental cost per B-1 flying hour than the system's average CPFH. Of course, we only observe B-1 fleet flying hours as low as 19,500 (for 2006) in any year, so it is a considerably out-of-sample extrapolation to estimate the system's fixed costs.

Also, this B-1 analysis is hampered by its small sample size—i.e., 11 years of data. We want to assess the general or most typical relationship between costs and flying hours across many systems. We could undertake Figure 3.1-type estimations for each MD, but each estimation would have a small sample size.

To assess a more general relationship between costs and FH totals, we chose a different path. In particular, we estimated an ordinary least-squares regression of the form

$$\ln(Cost_{it}) = a_i + b * \ln(FH_{it}) + c * Year_t + \varepsilon_{it},$$

Table 3.1
Example Dataset for the B-1

FY	MD	Flying Hours	TAI	CAIG Total	CAIG 2.0 Total
1996	B001	26,452.1	95.3	$1,096,323,121	$549,137,504
1997	B001	24,750.7	95.0	$1,014,964,289	$528,811,108
1998	B001	23,737.4	93.4	$1,118,814,538	$569,681,297
1999	B001	22,883.1	93.0	$1,067,326,050	$537,340,121
2000	B001	24,646.4	93.3	$1,154,331,456	$605,092,076
2001	B001	24,570.8	93.0	$1,192,743,820	$596,473,246
2002	B001	25,970.5	90.8	$1,318,086,915	$636,896,573
2003	B001	20,832.9	71.0	$1,183,338,481	$557,509,516
2004	B001	27,463.7	67.3	$1,265,734,021	$651,886,881
2005	B001	21,208.8	68.0	$1,125,190,817	$609,959,601
2006	B001	19,517.7	67.9	$1,073,315,770	$529,350,205

SOURCES: REMIS and AFTOC.

Figure 3.1
Linear Regression of B-1 Flying Hours–Expenditures

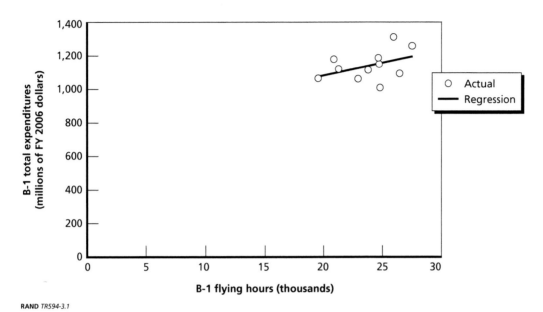

RAND *TR594-3.1*

where each i is an MD and each t is a year. In this estimation structure, each MD gets its own intercept ($\hat{a}_i$) but there is a common $\hat{b}$ that is the most typical empirical relationship between the natural log of flying hours and the natural log of costs. Intuitively, $\hat{b}$ is the estimated elasticity of cost with respect to flying hours. If flying hours increase 1 percent, on average, costs increase $\hat{b}$ percent.

The $c^{*}Year_t$ term is included to provide some assurance that we are not suffering from omitted variable bias by ignoring time trends, which could include, but are not limited to, age effects. For instance, if one felt there were shortcomings in AFTOC's translation of annual, nominal costs into real terms, this formulation would address such a concern. $Year_t$ is a dummy variable with the value 1 in FY t and 0 otherwise.[1]

In Figure 3.2, we illustrate the differences implied by different levels of $\hat{b}$.

If $\hat{b} > 1$, costs increase disproportionately as flying hours increase. If $\hat{b} = 1$, costs grow in proportion to flying hours. If $\hat{b} < 1$, costs do not grow in proportion to flying hours.

Figure 3.1's regression results are consistent with $\hat{b} < 1$. Indeed, if one runs a regression of $\ln(Cost_t)$ on $a + b^{*}\ln(FH_t)$ for the B – 1, one estimates that $\hat{b} = 0.28$, suggesting that a 10 percent increase in flying hours would increase total O&S spending by about 3 percent.

In Chapter Four, we run a series of $\ln(Cost_{it}) = a_i + b^{*}\ln(FH_{it}) + c^{*}Year_t + \varepsilon_{it}$ regressions using different categories of costs. In many cases, we find $\hat{b}$ values less than one, consistent with aircraft operation being characterized by sizable fixed costs and incremental costs less than the average CPFH.

---

[1] An alternative parameterization would be to include average fleet age as an independent variable. We undertook a number of such estimations, often deriving an "age effect" larger than we considered to be plausible. We ascertained the average fleet age was actually a proxy for an overall time trend, so we switched to the current structure with FY dummy variables. Our results were not greatly affected by this model structure alteration.

Figure 3.2
Implications of Different Values of $\hat{b}$

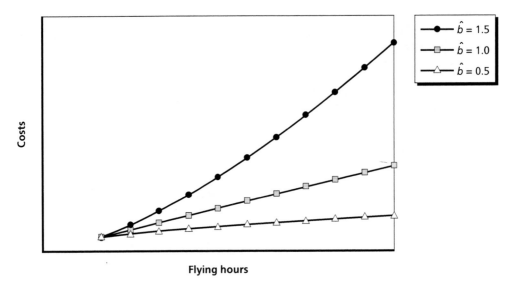

RAND *TR594-3.2*

CHAPTER FOUR

Estimation Results

In this chapter, we present a series of estimations with the natural log of various AFTOC MD-level annual expenditure totals as the dependent variables and the natural log of flying hours, FY dummy variables, and MD dummy variables as the independent variables.

We start with the highest-level regression with the natural log of total expenditures across all CAIG categories as the dependent variable. Table 4.1 presents the result.

Table 4.1's key result is its Ln(FH), $\hat{b}$, coefficient estimate of 0.56489. This result suggests total spending, on average, increases about 6 percent if an MD's flying hours increase 10 percent. This result is consistent with fixed costs and inconsistent with naïve application of CPFH.

The FY dummy variables (FY 1996–FY 2006) do not appear to be important with respect to $\hat{b}$, the parameter of central interest. All FY coefficients are measured relative to the omitted year, 1996. We reran Table 4.1's estimation removing the FY dummies. The resultant $\hat{b}$ estimate was 0.55885, down trivially from Table 4.1's coefficient estimate of 0.56489.

In terms of the MD dummy variables (in Table 4.1, the rows for the A-10 through the WC-130), all the coefficients are measured relative to the omitted MD, the C-130. It is important for estimation purposes that these MD-specific dichotomous independent variables are included, but we do not think they have great policy importance. We omit the FY and MD dummy variable coefficients from most of our subsequent regression displays.

While Table 4.1 covers all expenditures, it is of greater interest to us how the Ln(FH) coefficient varies across different level-one and level-two CAIG categories.

To examine level-one CAIG elements, we ran seven different log-log regressions, with the natural log of each CAIG level-one expenditure total as the dependent variables. Table 4.2 summarizes these additional seven regressions. We show the respective dependent variables on the left side of Table 4.2. In the middle of the table, we display the Ln(FH) coefficient estimates and their associated standard errors. The two right columns display two standard errors on each side of the Ln(FH) coefficient estimate, a traditional 95 percent confidence interval.

Each of the CAIG categories has an Ln(FH), $\hat{b}$, coefficient statistically significantly less than 1.0 except CAIG 3.0, intermediate maintenance, and CAIG 4.0, depot maintenance. CAIG 3.0 is a very small dollar category of little interest to us.

Much more important, the large CAIG 4.0 elasticity estimate is surprising. Many depot maintenance tasks are performed on a calendar, not flying-hour, basis, so we would expect a depot maintenance flying hour–spending elasticity less than, not greater than, 1.0.

Figure 4.1 shows the relationship, for all aircraft, between CAIG 4.0 expenditures and fleet-wide flying hours between FY 1996 and FY 2006. CAIG 4.0 expenditures increased nearly 80 percent, in real terms, between FY 2000 and FY 2003. Over the same period,

Table 4.1
Total Cost Regression Results

Observations	361
F (44,316)	386.24
Prob > F	0.0000
R-squared	0.9817
Root MSE	0.2136
Dependent variable	Ln (total costs all CAIG categories)

Independent Variable	Coefficient	SE	t	P > |t|
Ln(FH)	0.56489	0.02959	19.09	0.0000
Constant	14.43492	0.37324	38.67	0.0000
FY 1996	(omitted)			
FY 1997	−0.02784	0.05390	−0.52	0.6059
FY 1998	−0.03138	0.05389	−0.58	0.5608
FY 1999	0.10094	0.05353	1.89	0.0602
FY 2000	0.17477	0.05353	3.26	0.0012
FY 2001	0.17884	0.05323	3.36	0.0009
FY 2002	0.23362	0.05330	4.38	0.0000
FY 2003	0.30490	0.05323	5.73	0.0000
FY 2004	0.36950	0.05365	6.89	0.0000
FY 2005	0.44937	0.05373	8.36	0.0000
FY 2006	0.44993	0.05415	8.31	0.0000
A-10	−0.29103	0.09318	−3.12	0.0020
AC-130	−0.52226	0.13172	−3.96	0.0000
AT-38	−1.69097	0.12565	−13.46	0.0000
B-1	0.53173	0.11310	4.70	0.0000
B-2	0.35277	0.14417	2.45	0.0150
B-52	0.21250	0.11263	1.89	0.0601
C-5	0.41767	0.09805	4.26	0.0000
C-9	−0.97498	0.12819	−7.61	0.0000
C-17	−0.33174	0.09678	−3.43	0.0007
C-20	−1.62812	0.14312	−11.38	0.0000
C-21	−2.01747	0.10237	−19.71	0.0000
C-26	−2.62207	0.14127	−18.56	0.0000
C-37	−1.90738	0.15397	−12.39	0.0008
C-130	(omitted)			
C-141	−0.08574	0.10478	−0.82	0.4138
E-3	−0.03788	0.11557	−0.33	0.7433
E-8	−0.16522	0.14713	−1.12	0.2623
EC-130	−1.52285	0.14767	−10.31	0.0000
F-15	0.54268	0.09129	5.94	0.0000
F-16	0.38934	0.09191	4.24	0.0000
F-117	−0.24474	0.12557	−1.95	0.0522
HC-130	−1.02590	0.13041	−7.87	0.0000

Table 4.1—Continued

Independent Variable	Coefficient	SE	t	P > |t|
KC-10	−0.42899	0.10010	−4.29	0.0000
KC-135	0.19261	0.09116	2.11	0.0354
LC-130	−1.77529	0.14971	−11.86	0.0000
MC-130	−0.38474	0.11246	−3.42	0.0007
RC-135	−0.21965	0.12703	−1.73	0.0848
T-1	−2.10448	0.09538	−22.06	0.0000
T-6	−2.53871	0.12000	−21.16	0.0000
T-37	−1.54309	0.09168	−16.83	0.0000
T-38	−1.20700	0.09294	−12.99	0.0000
T-43	−2.19825	0.14649	−15.01	0.0000
U-2	−0.14472	0.12920	−1.12	0.2635
WC-130	−1.58099	0.16052	−9.85	0.0000

NOTE: FY 1996 is omitted (as is the C-130) so that all FY (MD) coefficients can be measured relative to the omitted FY (MD).

Table 4.2
Different CAIG Level-One Ln(FH) Regression Coefficient Estimates

Dependent Variable	Category	Ln(FH) Coefficient			
		Estimate	SE	Estimate − 2SE	Estimate + 2SE
Ln(total spending)	Total spending	0.56489	0.02959	0.50571	0.62407
Ln(CAIG 1.0)	Mission personnel	0.52008	0.04614	0.42780	0.61236
Ln(CAIG 2.0)	Unit-level consumption	0.87939	0.03150	0.81639	0.94239
Ln(CAIG 3.0)	Intermediate maintenance	0.40890	0.33402	−0.25914	1.07694
Ln(CAIG 4.0)	Depot maintenance	1.42685	0.15468	1.11749	1.73621
Ln(CAIG 5.0)	Contractor support	0.40096	0.11412	0.17272	0.62920
Ln(CAIG 6.0)	Sustaining support	−0.13004	0.13538	−0.40080	0.14072
Ln(CAIG 7.0)	Indirect support	0.32007	0.05423	0.21161	0.42853

fleet-wide flying hours increased about 20 percent, so the regression estimation attributes the marked CAIG 4.0 expenditure increase to the comparatively moderate flying-hour increase.

This result is almost certainly spurious. Figure 4.2 reprises Figure 4.1's data, but also presents expenditures for CAIG 2.3, depot-level reparables. Whereas expenditures for CAIG 4.0, depot maintenance, increased nearly 80 percent between FY 2000 and FY 2003, CAIG 2.3 expenditures jumped about 20 percent from FY 1998 to FY 1999, peaked in 2002, and have since declined in real terms.

A regression of CAIG 2.3 expenditures on flying hours estimates a $\hat{b}$ value of 0.09175, not significantly different than zero. This result is not surprising since CAIG 2.3 expenditures

Figure 4.1
The Relationship Between Fleet-Wide Flying Hours and CAIG 4.0 Expenditures, FY 1996 to FY 2006

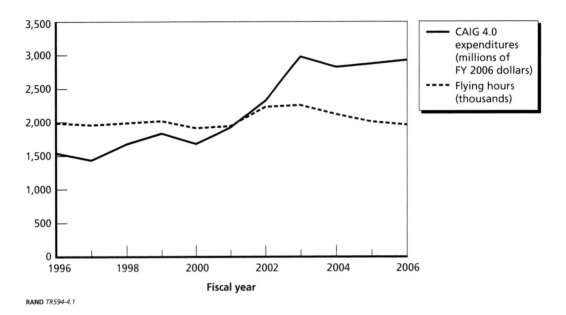

RAND *TR594-4.1*

Figure 4.2
The Relationship Between Fleet-Wide Flying Hours, CAIG 4.0 Expenditures, and CAIG 2.3 Expenditures, FY 1996 to FY 2006

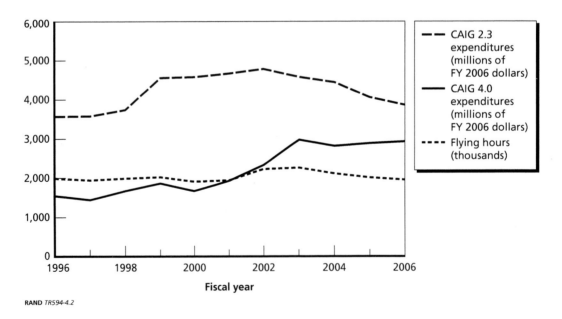

RAND *TR594-4.2*

were virtually unchanged, in real terms, between FY 2000 and FY 2003, a period during which fleet-wide flying hours increased markedly.

We do not understand what has occurred with these depot-related cost categories. The data make it appear as if the Air Force transferred some costs from CAIG 2.3 (depot-level repa-rables) to CAIG 4.0 (depot maintenance) during this period. But no expert we have contacted has corroborated that something like that happened.

Irrespective of the cause of the observed phenomenon, we do not think it is useful for our analysis to separate CAIG 2.3 and CAIG 4.0 expenditures.

Figure 4.3 shows a plot of the sum of CAIG 2.3 and CAIG 4.0 constant dollar expenditures (the left-hand y-axis) against fleet-wide flying hours (the right-hand y-axis). Total depot expenditures are still escalating over time, but the pattern is much more similar to the flying-hour pattern.

We then ran the Ln(FH) regression with Ln(CAIG 2.3 + CAIG 4.0) as the dependent variable. The Ln(FH) coefficient estimate, $\hat{b}$, was 0.22280, with a standard error of 0.13253. Obviously, this is a vastly smaller flying hour–spending elasticity estimate than we found for CAIG 4.0 alone. Indeed, one cannot reject a null hypothesis that flying hours have no effect on depot maintenance expenditures in this estimation. We aggregate 2.3 and 4.0 expenditures in all of our subsequent analyses.

Level-Two Estimations

We undertook the same types of log-log regressions for the level-two expenditure categories. We excluded, however, 2.3 (depot-level reparables) and the 4.0 (depot maintenance) categories based on our belief that these categories should be aggregated. Reprising Table 4.2's format, Table 4.3 shows the different two-level Ln(FH) regression coefficients. Not all of these regressions cover as many aircraft as Table 4.1's since some of these cost categories had no expenditures for some aircraft. Such observations were omitted from the estimations. As in Table 4.2, each row displays a different dependent variable, then the Ln(FH) coefficient estimate, its standard error, and plus and minus two standard errors around the Ln(FH) coefficient estimate.

Figure 4.3
The Relationship Between Fleet-Wide Flying Hours and CAIG 2.3 and CAIG 4.0 Combined Expenditures, FY 1996 to FY 2006

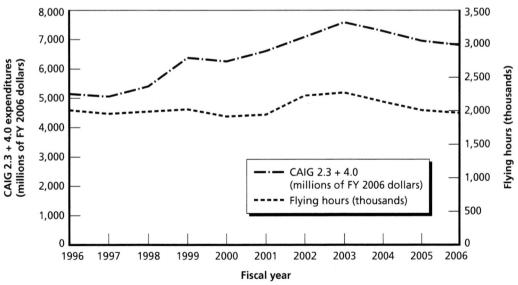

Table 4.3
Different CAIG Level-Two Ln(FH) Regression Coefficient Estimates

Dependent Variable	Category	Ln(FH) Coefficient			
		Estimate	SE	Estimate – 2SE	Estimate + 2SE
Ln(CAIG 1.2)	Maintenance personnel	0.58097	0.06548	0.45001	0.71193
Ln(CAIG 1.3)	Other mission personnel	0.44522	0.05396	0.33730	0.55314
Ln(CAIG 2.1)	POL/energy consumption	1.08181	0.06941	0.94299	1.22063
Ln(CAIG 2.2)	Consumables	0.30436	0.08791	0.12854	0.48018
Ln(CAIG 2.4)	Training munitions	0.52960	0.74110	–0.95260	2.01180
Ln(CAIG 2.5)	Other unit-level consumption	0.71092	0.11407	0.48278	0.93906
Ln(CAIGs 2.3 + 4.0)	Depot maintenance	0.22280	0.13253	–0.04226	0.48786
Ln(CAIG 5.2)	Contractor logistical support	0.49031	0.12320	0.24391	0.73671
Ln(CAIG 5.3)	Other contractor support	0.42796	0.29453	–0.16110	1.01702
Ln(CAIG 6.1)	Support equipment replacement	0.16271	0.15680	–0.15089	0.47631
Ln(CAIG 6.3)	Other recurring investment	0.33128	0.17537	–0.01946	0.68202
Ln(CAIG 6.4)	Sustaining engineering support	–0.12600	0.15503	–0.43606	0.18406
Ln(CAIG 6.5)	Software maintenance	0.71944	0.38351	–0.04758	1.48646
Ln(CAIG 6.6)	Simulator operations	–0.46121	0.99210	–2.44510	1.52299
Ln(CAIG 7.1)	Personnel support	0.60759	0.07587	0.45585	0.75933
Ln(CAIG 7.2)	Installation support	0.29373	0.05945	0.17483	0.41263

Table 4.4 then notes for which categories we can reject $\hat{b}$ values of 0 (no relation with flying hours) and 1 (proportional relationship to flying hours) at the 95 percent confidence level.

For eight (1.1, 1.2, 1.3, 2.2, 2.5, 5.2, 7.1, and 7.2) of the 16 CAIG categories, the regression analysis rejects both $\hat{b} = 0$ and $\hat{b} = 1$. We label such cases "partially variable," i.e., they apparently have fixed costs that imply the category's costs do not grow in proportion to flying hours, yet there appears to be some relationship to flying hours.

In seven (2.4, 5.3, 6.1, 6.3, 6.4, 6.5, and 6.6) of the categories, we cannot reject $\hat{b} = 0$, i.e., a hypothesis that there is no relationship between the category's expenditure level and flying hours. The hybrid 2.3 + 4.0 depot maintenance category also fails to reject $\hat{b} = 0$.

Table 4.4
Different CAIG Level-Two Ln(FH) Regression Coefficient Hypothesis Tests

Dependent Variable	Category	Reject $\hat{b} = 0$?	Reject $\hat{b} = 1$?
Ln(CAIG 1.1)	Operations personnel	Yes	Yes
Ln(CAIG 1.2)	Maintenance personnel	Yes	Yes
Ln(CAIG 1.3)	Other mission personnel	Yes	Yes
Ln(CAIG 2.1)	POL/energy consumption	Yes	No
Ln(CAIG 2.2)	Consumables	Yes	Yes
Ln(CAIG 2.4)	Training munitions	No	No
Ln(CAIG 2.5)	Other unit-level consumption	Yes	Yes
Ln(CAIG 2.3 + 4.0)	Depot maintenance	No	Yes
Ln(CAIG 5.2)	Contractor logistical support	Yes	Yes
Ln(CAIG 5.3)	Other contractor support	No	No
Ln(CAIG 6.1)	Support equipment replacement	No	Yes
Ln(CAIG 6.3)	Other recurring investment	No	Yes
Ln(CAIG 6.4)	Sustaining engineering support	No	Yes
Ln(CAIG 6.5)	Software maintenance	No	No
Ln(CAIG 6.6)	Simulator operations	No	No
Ln(CAIG 7.1)	Personnel support	Yes	Yes
Ln(CAIG 7.2)	Installation support	Yes	Yes

Categories 2.4 (training munitions), 5.3 (other contractor support), 6.5 (software maintenance), and 6.6 (simulator operations) are so imprecisely estimated with these data that we also cannot reject $\hat{b} = 1$, costs rising in proportion to flying hours. The results for 6.1 (support equipment replacement), 6.3 (other recurring investment), 6.4 (sustaining engineering support), and 2.3 + 4.0 (depot maintenance) seem more solid because one can reject costs growing in proportion to flying hours for those categories.

Category 2.1 (POL/energy consumption) is the only one for which we can reject $\hat{b} = 0$ without rejecting $\hat{b} = 1$. Of course, this is not a surprising result: We know energy consumption is highly correlated with flying hours. The only real source of noise in POL expenditure data is fluctuation in energy prices over time. But, certainly, a priori, if any category should exhibit high correlation with flying hours, it would be energy expenditures. This hypothesis is borne out in the data.

Costs Vary by Tail?

Our analyses to this point have had the natural log of flying hours as the independent variable of greatest interest. As shown in Figures 1.1 and 1.2, however, SAF/FM categorizes some types of expenses as variable by tail, not by flying hour.

There is an analytical challenge with inclusion of both flying hours and fleet size as independent variables. In a peacetime steady-state, we expect a high correlation between the number of aircraft of a given MD and the total number of hours flown by that MD. Indeed, if every Air Force airplane flew say, one hour per day, fleet size and fleet flying hours would be perfectly correlated and it would be impossible to distinguish fleet size effects from fleet flying hour effects.

However, as noted by Cook, Ausink, and Roll (2005), the Air Force's traditionally stable relationship between fleet size and flying hours was perturbed after the events of September 11, 2001. In particular, cargo, and to a lesser extent bomber and tanker, aircraft saw sharp increases in their flying hours per tail. Consequently, there was a disconnection between flying hours and fleet sizes and, therefore, an opportunity to assess the separate influences of flying-hour and fleet-size changes.

Table 4.5 presents the natural log of total cost regression with both Ln(FH) and Ln(TAI) included as independent variables. Table 4.5's structure mimics that of Table 4.1's with the variables of central interest at the top, followed by the FY dummy variables, and, finally, the MD dummy variables.

Table 4.5 suggests both flying hours and fleet sizes have partial effects on total costs, with both coefficient estimates being significantly greater than zero but also significantly less than one.

Even with the post-9/11 surge in some fleets' flying hours per tail, the Pearson correlation between Ln(FH) and Ln(TAI) in our data is 0.95621. Given such a large correlation, it is somewhat remarkable that Table 4.5 finds both variables to be statistically significant. One surmises that such an estimation would have been infeasible before the post-9/11 flying-hour surge.

Of course, our central interest is to look at specific categories of expenditures and to assess their flying-hour and fleet-size sensitivities. We therefore reran our level-two regressions with both Ln(FH) and Ln(TAI) as independent variables. Table 4.6 presents the results. Table 4.6 displays the respective expenditure categories as dependent variables, then the Ln(FH) coefficient with its standard error, and finally the Ln(TAI) coefficient with its standard error. (Because of table width constraints, we omit display of the 95 percent confidence intervals presented in Tables 4.2 and 4.3)

Table 4.6's coefficients are difficult to interpret by themselves. In an effort to bring more coherence to these results, Table 4.7 puts some of Table 4.6's coefficients into different categories.

Table 4.7 finds considerably different tendencies across different CAIG elements. Category 7.2 (installation support) costs appear to be fixed, both with respect to the number of aircraft and the number of flying hours. This result accords with SAF/FM's categorization of 7.2, as shown in Figure 1.2.

Category 2.1 (POL/energy consumption) appears to vary in proportion to the number of flying hours, but not with the number of aircraft. SAF/FM categorized 2.1 as varying with flying hours so this finding also agrees with Figure 1.2.

Table 4.7 finds 1.3 (other mission personnel) to have a partial relationship with the number of aircraft, but no apparent relationship to flying hours. SAF/FM categorized 1.3 as variable with the number of aircraft, so we would suggest an alternative fixed-and-variable function describing the relationship between the number of aircraft and expenditures for other mission personnel.

Table 4.7 finds 1.1 (operations personnel) to be partially varying with flying hours but not with the number of aircraft. This is different than SAF/FM's view that 1.1 expenses vary with the number of aircraft.

Table 4.5
Total Cost Regression Results with Flying Hours and Fleet Size as Independent Variables

Observations	361
F (45,315)	395.62
Prob > F	0.0000
R-squared	0.9826
Root MSE	0.20878
Dependent variable	Ln(total costs, all CAIG categories)

Independent Variable	Coefficient	SE	t	P > |t|
Ln(FH)	0.29167	0.07469	3.91	0.0001
Ln(TAI)	0.36327	0.09156	3.97	0.0000
Constant	15.53924	0.45887	33.86	0.0000
FY 1996	(omitted)			
FY 1997	−0.03781	0.05274	−0.72	0.4740
FY 1998	−0.03868	0.05270	−0.73	0.4635
FY 1999	0.10414	0.05233	1.99	0.0474
FY 2000	0.16597	0.05237	3.17	0.0017
FY 2001	0.17066	0.05207	3.28	0.0012
FY 2002	0.25664	0.05242	4.90	0.0000
FY 2003	0.32726	0.05234	6.25	0.0000
FY 2004	0.38106	0.05252	7.26	0.0000
FY 2005	0.44685	0.05253	8.51	0.0000
FY 2006	0.45343	0.05293	8.57	0.0000
A-10	−0.35095	0.09233	−3.80	0.0002
AC-130	−0.24437	0.14657	−1.67	0.0964
AT-38	−1.70111	0.12285	−13.85	0.0000
B-1	0.57100	0.11099	5.14	0.0000
B-2	0.50699	0.14618	3.47	0.0006
B-52	0.21697	0.11009	1.97	0.0496
C-5	0.60184	0.10648	5.65	0.0000
C-9	−0.49491	0.17419	−2.84	0.0048
C-17	0.10250	0.14466	0.71	0.4792
C-20	−1.28025	0.16509	−7.75	0.0000
C-21	−1.76467	0.11862	−14.88	0.0000
C-26	−2.25608	0.16606	−13.59	0.0000
C-37	−1.34961	0.20594	−6.55	0.0000
C-130	(omitted)			
C-141	0.07999	0.11061	0.72	0.4701

Table 4.5—Continued

Independent Variable	Coefficient	SE	t	P > |t|
E-3	0.30567	0.14233	2.15	0.0325
E-8	0.29295	0.18444	1.59	0.1132
EC-130	–1.15565	0.17146	–6.74	0.0000
F-15	0.35346	0.10118	3.49	0.0005
F-16	0.13693	0.11009	1.24	0.2145
F-117	–0.21247	0.12301	–1.73	0.0851
HC-130	–0.87335	0.13315	–6.56	0.0000
KC-10	–0.02809	0.14065	–0.20	0.8418
KC-135	0.14043	0.09007	1.56	0.1200
LC-130	–1.35185	0.18112	–7.46	0.0000
MC-130	–0.23123	0.11653	–1.98	0.0481
RC-135	0.19351	0.16205	1.19	0.2333
T-1	–1.97659	0.09865	–20.04	0.0000
T-6	–2.40123	0.12230	–19.63	0.0000
T-37	–1.54779	0.08962	–17.27	0.0000
T-38	–1.33787	0.09665	–13.84	0.0000
T-43	–1.82695	0.17106	–10.68	0.0000
U-2	0.12829	0.14382	0.89	0.3730
WC-130	–1.37342	0.16540	–8.30	0.0000

NOTE: FY 1996 is omitted (as is the C-130) so that all FY (MD) coefficients can be measured relative to the omitted FY (MD).

Table 4.7 finds 1.2 (maintenance personnel), 2.2 (consumables/repair parts), 2.5 (other unit-level consumption), 5.2 (contractor logistics support), 6.1 (support equipment replacement), and 7.1 (personnel support) to vary in proportion to the number of aircraft, but not with flying hours. SAF/FM categorized 1.2 as varying with the number of aircraft, but asserted 2.2, 2.5, and 5.2 to vary with flying hours, while 6.1 and 7.1 were fixed.

We omitted our newly agglomerated depot-maintenance category (the sum of the 2.3 and the 4.0 categories) from Table 4.7. This new category shows no significant relationship to the number of flying hours. The relationship between expenditures and the number of tails is highly imprecise; one cannot reject either no correlation or one-to-one correlation with the number of tails. This estimation technique has not provided great insight about this category. We urge further research into the drivers of aircraft depot maintenance costs. The analysis technique presented in this report has proven incapable of shedding meaningful insight on depot maintenance costs.

Table 4.7 also omits 2.4 (training munitions), 3.0 (intermediate maintenance), 5.3 (other contractor support), 6.3 (other recurring investments), 6.4 (sustaining equipment support), 6.5 (software maintenance), and 6.6 (simulator operations). The results for these categories were not coherent. Categories 3.0, 5.3, 6.5, and 6.6 had results in which the coefficient estimates were so imprecise that values of 0 and 1 could not be rejected for either the Ln(FH) or Ln(TAI)

Table 4.6
Ln(FH) and Ln(TAI) Regression Coefficient Estimates

Dependent Variable	Category	Ln(FH) Coefficient Estimate	Ln(FH) Coefficient SE	Ln(TAI) Coefficient Estimate	Ln(TAI) Coefficient SE
Ln(Total Spending)	Total spending	0.29167	0.07469	0.36327	0.09156
Ln(CAIG 1.1)	Operations personnel	0.41980	0.12432	0.25398	0.15195
Ln(CAIG 1.2)	Maintenance personnel	0.05897	0.16736	0.69721	0.20631
Ln(CAIG 1.3)	Other mission personnel	0.05999	0.13773	0.51818	0.17085
Ln(CAIG 2.1)	POL/energy consumption	1.21080	0.17935	−0.17149	0.21986
Ln(CAIG 2.2)	Consumables	−0.37549	0.19974	0.94268	0.25001
Ln(CAIG 2.3 + 4.0)	Depot maintenance	−0.29156	0.31153	0.69318	0.38030
Ln(CAIG 2.4)	Training munitions	−3.31964	1.38154	5.32346	1.64418
Ln(CAIG 2.5)	Other unit-level consumption	0.06027	0.27109	0.89070	0.33739
Ln(CAIG 3.0)	Intermediate maintenance	−0.06049	1.08949	0.50954	1.12513
Ln(CAIG 5.2)	Contractor logistical support	−0.18977	0.29930	0.91421	0.36729
Ln(CAIG 5.3)	Other contractor support	0.38646	0.65063	0.05653	0.78977
Ln(CAIG 6.1)	Support equipment replacement	−0.70455	0.37310	1.18644	0.46409
Ln(CAIG 6.3)	Other recurring investment	0.54754	0.37253	−0.32287	0.49047
Ln(CAIG 6.4)	Sustaining engineering support	−0.34496	0.34555	0.28035	0.41420
Ln(CAIG 6.5)	Software maintenance	0.79336	0.68207	−0.11984	0.91295
Ln(CAIG 6.6)	Simulator operations	−0.92879	1.47905	0.91514	2.13541
Ln(CAIG 7.1)	Personnel support	0.06414	0.19339	0.72256	0.23707
Ln(CAIG 7.2)	Installation support	0.21996	0.15369	0.09809	0.18840

Table 4.7
Summary of Ln(FH) and Ln(TAI) Regression Coefficient Estimates

		Ln(FH) Coefficient No Relationship to FH	Ln(FH) Coefficient Partial Relationship to FH	Ln(FH) Coefficient Proportional to FH
Ln(TAI) Coefficient	No relationship to TAI	7.2	1.1	2.1
	Partial relationship to TAI	1.3	Total spending	
	Proportional to TAI	1.2, 2.2, 2.5, 5.2, 6.1, 7.1		

coefficient. For 6.3, one can reject the Ln(TAI) coefficient being 1, but the Ln(FH) coefficient could be 0 or 1. For 6.4, the opposite was true: One could reject the Ln(FH) coefficient being 1, but the Ln(TAI) could be 0 or 1. The 2.4 coefficient estimates were especially peculiar, with a statistically significantly negative coefficient on Ln(FH) and an Ln(TAI) coefficient statistically significantly greater than 1.

An Alternative Categorization Scheme

In Figures 1.1 and 1.2, we presented SAF/FM's breakout of costs by CAIG category. In Figure 4.4, we present an alternative portrayal based on this chapter's analysis.

More important than the number of CAIG categories in each box is the relative importance of each category. In Figure 4.5, we compare SAF/FM's to our dichotomy as percentages of total FY 1996–FY 2006 constant dollar expenditures.

Given that we have introduced new categorizations, it is not surprising that we find that SAF/FM has a greater percentage of costs allocated to both "TAI variable" and, especially, "FH variable." The large "indeterminate" category is driven by the ambiguous results for the depot-maintenance expenditure category (2.3 plus 4.0).

The second most important new category is "partial FH," our finding that operations personnel costs (1.1) appear to vary with flying hours, though not proportionally. This is a more subtle possibility than SAF/FM's simpler categorization admitted.

Figure 4.4
Alternative Air Force Expenditure Categorization Scheme

Variable cost per flying hour
2.1 POL/energy consumption

Variable cost per TAI
1.2 Maintenance personnel
2.2 Consumables/repair parts
2.5 Other unit-level consumption
5.2 Contractor logistics support
6.1 Support equipment
7.1 Personnel support

Partially variable cost per flying hour
1.1 Operations personnel

Partially variable cost per TAI
1.3 Other mission personnel

Fixed costs
7.2 Installation support

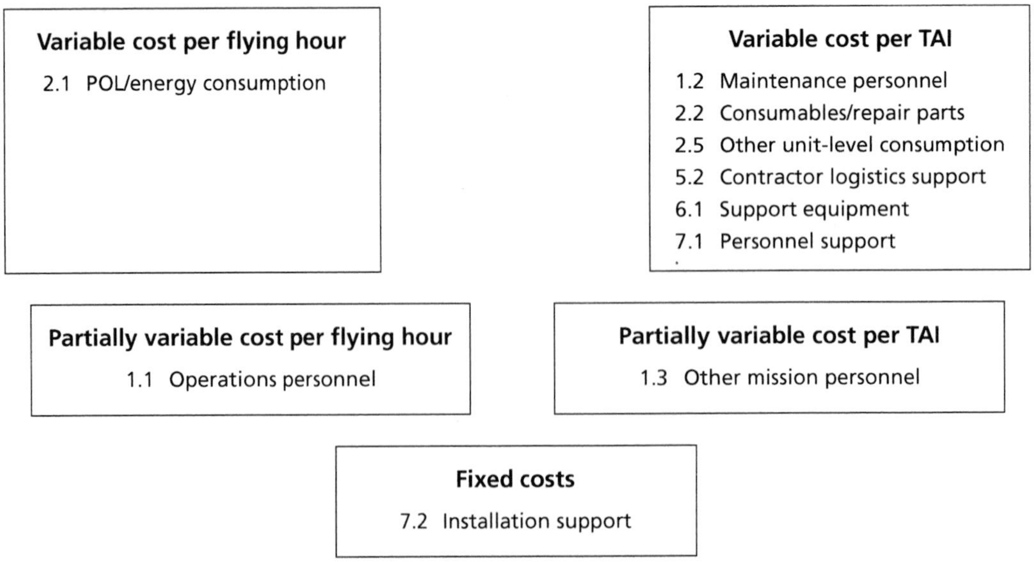

NOTE: We cannot determine 2.3 + 4.0 (depot maintenance), 2.4 (training munitions), 3.0 (intermediate maintenance), 5.3 (other contract support), 6.3 (other recurring investment), 6.4 (sustaining equipment support), 6.5 (software maintenance), and 6.6 (simulator operations).
RAND *TR594-4.4*

Figure 4.5
The Dollar Value Weighting of Alternative Categorizations

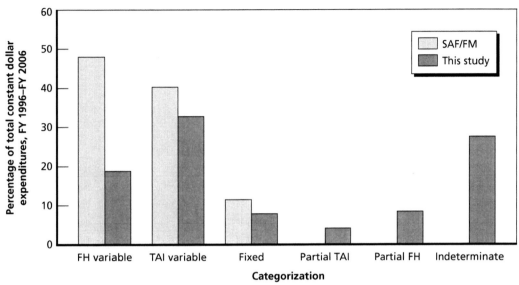

CHAPTER FIVE

Policy Implications

A major objective of this research is to improve the Air Force's budgeting process. When planned flying hours or fleet sizes change, budgets should be adjusted appropriately, neither excessively nor insufficiently.

As discussed throughout this report, SAF/FM currently categorizes types of expenditures as varying with flying hours, varying with fleet sizes, or fixed, i.e., varying with neither flying hours nor fleet size.

Our analysis of FY 1996–2006 expenditure and flying-hour data suggests possible improvements to SAF/FM's approach. In particular, the "varying-with-flying-hours" and "varying-with-fleet-size" categorizations are too simplistic. In fact, some expenditure categories appear to exhibit *fixed-and-variable* characteristics, e.g., there is a baseline level of costs in the category then costs increase with flying hours or fleet size, but not proportionally.

A simplified pedagogical example illustrates our concern with SAF/FM's current approach. SAF/FM views eight CAIG categories as varying with flying hours. One of those eight is 2.3, depot-level reparables, which we think should be merged with 4.0, depot maintenance. We also omit 2.4, training munitions, because the Air Force Cost and Performance (AFCAP) tool does not provide future budget information about this category.

The remaining six categories are presented in Table 5.1 along with their 2006 Air Force–wide expenditure levels. For each category, we display the category's actual FY 2006 expenditure level, the category's projected FY 2007 expenditure level if expenditures fell in proportion

Table 5.1
SAF/FM's Variable with Flying Hour Categories

CAIG	Category Title	FY 2006 Expenditures (millions)	FH-Proportional FY 2007 Expenditures (millions)	Table 4.3 Ln(FH) Coefficient Estimate	Elasticity-Adjusted FY 2007 Expenditures (millions)
2.1	POL/energy consumption	$5,733.7	$5,165.6	1.08181	$5,121.7
2.2	Consumables	$1,131.6	$1,019.5	0.30436	$1,096.3
2.5	Other unit-level consumption	$545.1	$491.1	0.71092	$506.2
3.0	Intermediate maintenance	$0.5	$0.5	0.40096[a]	$0.5
5.2	Contractor logistical support	$4,204.5	$3,787.9	0.49031	$3,994.8
5.3	Other contractor support	$121.6	$109.5	0.42796	$116.3
	Total	$11,737.0	$10,574.1		$10,835.7

[a] This coefficient estimate is from Table 4.2.

to planned flying hours, the Table 4.3 Ln(FH) coefficient estimate for that category, and the category's projected FY 2007 expenditure level if Table 4.3's flying-hour elasticity was valid.

AFCAP projected a fleet-wide decrease in flying hours from 2,113,643 in FY 2006 to 1,904,229 in FY 2007. Though actual budgeting would be done on a system-by-system basis, Table 5.1's fourth column simply reduces each category's constant dollar expenditure level proportionally. This is the logical consequence of these categories' expenditures being variable with flying hours.

Table 4.3, however, shows different estimated flying hour elasticities for these categories. All except 2.1 are significantly less than 1.0, so we do not expect their expenditures to fall in proportion to flying hours. The right-most column of Table 5.1 shows alternative projected FY 2007 expenditure levels. For all except 2.1, they are greater than those projected by assuming flying-hour proportionality.[1]

If flying hours are falling, categorizing expenditures as variable that are actually partially fixed will lead to excessive budget cuts. If flying hours are rising, categorizing expenditures as variable that are actually partially fixed will lead to excessive budget increases.

---

[1] One could equally well use Table 4.6's regression coefficients, except we do not have a projection of the change in fleet size.

Bibliography

Armstrong, Patrick D., *Developing an Aggregate Marginal Cost per Flying Hour Model for Air Force's F-15 Fighter Aircraft*, thesis, Wright-Patterson AFB, Ohio: Air Force Institute of Technology, March 2006.

Cook, Cynthia R., John A. Ausink, and Charles Robert Roll Jr., *Rethinking How the Air Force Views Sustainment Surge*, Santa Monica, Calif.: RAND Corporation, MG-372-AF, 2005. As of October 27, 2008: http://www.rand.org/pubs/monographs/MG372/

Hawkes, Eric M., *Predicting the Cost per Flying Hour for the F-16 Using Programmatic and Operational Variables*, thesis, Wright-Patterson AFB, Ohio: Air Force Institute of Technology, June 2005.

Hildebrandt, Gregory G., and Man-bing Sze, *An Estimation of USAF Aircraft Operating and Support Cost Relations*, Santa Monica, Calif.: RAND Corporation, N-3062-ACQ, 1990. As of October 27, 2008: http://www.rand.org/pubs/notes/N3062/

Laubacher, Matthew E., *Analysis and Forecasting of Air Force Operating and Support Cost for Rotary Aircraft*, thesis, Wright-Patterson AFB, Ohio: Air Force Institute of Technology, March 2004.

Lies, Tom, and Larry Klapper, "Air Force Cost and Performance (AFCAP) Tool," briefing, presented at AFTOC [Air Force Total Ownership Cost] Users Conference, March 2007.

Office of the Secretary of Defense, Cost Analysis Improvement Group, *Operating and Support Cost-Estimating Guide*, Washington, D.C., May 1992. As of October 27, 2008: http://www.dtic.mil/pae/

Rose, Pat A. Jr., "Cost per Flying Hour Factors: A Background and Perspective of How They Are Developed and What They Do," *The Air Force Comptroller*, Vol. 31-1, No. 4-9, July 1997.

Secretary of the Air Force, Financial Management & Comptroller, *Logistics Cost Factors Description*, Air Force Instruction 65-503, November 2005.

Sherbrooke, Craig C., *Using Sorties vs. Flying Hours to Predict Aircraft Spares Demand*, McLean, Va.: Logistics Management Institute, Report AF501LN1, April 1997.

Slay, F. Michael, *Demand Forecasting*, McLean, Va.: Logistics Management Institute, Report AF401N2, August 1995.

U.S. Air Force, "U.S. Air Force Cost and Planning Factors," Air Force Instruction (AFI) 65-503, February 4, 1994.

———, Air Force Cost Analysis Improvement Group, *Cost per Flying Hour Process Guide*, November 1999.

U.S. Department of Defense (DoD), Cost Analysis Improvement Group (CAIG), DoD Directive 5000.04, August 16, 2006. As of October 27, 2008: http://www.dtic.mil/whs/directives/corres/pdf/500004p.pdf

Wallace, John M., Scott A. Houser, and David A. Lee, *A Physics Based Alternative to Cost-Per-Flying-Hour Models of Aircraft Consumption Costs*, McLean, Va.: Logistics Management Institute, Report AF909T1, August 2000.